Free Open Source Antivirus Software
Untuk Sistem Operasi Ubuntu
Linux Edisi Bahasa Inggris
Lite Version

by

Cyber Jannah Sakura

2020

Pendahuluan

All computer systems can suffer from malware, ransomware and viruses including BSD, Linux & Mac. Thankfully, very few viruses exist for Linux, so users typically do not install antivirus software. It is still recommended that Linux users have antivirus software installed on Linux systems that are on a network or that have files being transferred to the device. Some users may argue that antivirus software uses up too much resources. Thankfully, low-footprint software exists for Linux. To better understand antivirus programs, it may be beneficial to understand malware itself.

If you are running a SME business with a number of workstations, it might be a good idea to install an antivirus on the central computer that manages all the emails, data and traffic in your company. The best way to protect a system against viruses is to only download and install software from trusted sites and developers.

With the discontinuation of AVG Antivirus for Linux and the lack of security suite from big players such as Symantec Nortonor Intel McAfee, Linux Operating System are left with a few choices when it comes to OS security. Linux users are generally free from virus attack but bugs that enables a hacker to take over your linux system is out there, thus an antivirus with a good firewall is a must for your Ubuntu, Mint, Debian, Elementary, Fedora, CentOS and more.

1. Comodo Antivirus

Comodo Antivirus for Linux (CAVL) offers the same great virus protection as our Windows software with the added benefit of a fully configurable anti-spam system.

Featuring on-access and on-demand virus scanners, CAVL also uses cloud based behavior analysis of unknown files to provide unrivaled protection against zero-day malware. The powerful AV engine is complemented by a highly configurable mail gateway to filter spam and block email-borne threats. Features

- Proactive AV protection intercepts all known threats
- Automatic updates for the most up-to-date virus protection
- Includes scan scheduler, detailed event viewer and custom scan profiles
- Mail filter is compatible with Postfix, qmail, Sendmail and Exim MTA's
- Install and forget. No annoying false alarms, just solid virus protection.

Powerful anti-virus and email filtering software for Linux based computers.

- Detects and eliminates viruses from Linux based laptops and desktops
- Mail gateway blocks email-borne viruses and spam
- Features on-access, on-demand and cloud based scanning
- Virus definitions updated multiple times per day
- Simple to use: install and forget while Comodo Antivirus protects you in the background.

2. F-PROT for Linux Workstations

For home users using the Linux open-source operating system, we offer F-Prot Antivirus for Linux Workstations. F-PROT Antivirus for Linux Workstations utilizes the renowned F-PROT Antivirus scanning engine for primary scan but has in addition to that a system of internal heuristics devised to search for unknown viruses

F-PROT Antivirus for Linux was especially developed to effectively eradicate viruses threatening workstations running Linux. It provides full protection against macro viruses and other forms of malicious software – including Trojans. F-PROT for Linux Workstations features:

- Scans for over 2119958 known viruses and their variants
- Ability to perform scheduled scans when used with the cron utility
- Scans hard drives, CD-ROMS, diskettes, network drives, directories and specific files
- Scans for images of boot sector viruses, macro viruses and Trojan Horses

3. Clamav Free Antivirus

ClamAV is an open source (GPL) anti-virus engine used in a variety of situations including email scanning, web scanning, and end point security. It provides a number of utilities including a flexible and scalable multi-threaded daemon, a command line scanner and an advanced tool for automatic database updates. Features

- Command-line scanner
- Milter interface for sendmail
- Advanced database updater with support for scripted updates and digital signatures
- Virus database updated multiple times per day
- Built-in support for all standard mail file formats
- Built-in support for various archive formats, including Zip, RAR, Dmg, Tar, Gzip, Bzip2, OLE2, Cabinet, CHM, BinHex, SIS and others
- Built-in support for ELF executables and Portable Executable files packed with UPX, FSG, Petite, NsPack, wwpack32, MEW, Upack and obfuscated with SUE, Y0da Cryptor and others
- Built-in support for popular document formats including MS Office and MacOffice files, HTML, Flash, RTF and PDF

4. Sophos Linux Malware Scanner

The Sophos Antivirus engine effectively detects and cleans viruses, Trojans, and other malware. In addition to sophisticated detection-based on advanced heuristics, Sophos Antivirus for Linux uses Live Protection to look up suspicious files in real time via SophosLabs.

To prevent the Linux machine from becoming a distribution point, Sophos Antivirus for Linux also detects, blocks, and removes Windows, Mac, and Android malware. Features:

- Detects and blocks malware with on-access, on-demand, or scheduled scanning

- Excellent performance, low impact

5. F-Secure Linux Security

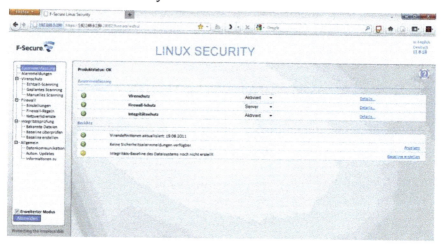

F-Secure is the most complete security software for Linux, it has ever feature from Firewall to Cloud Antivirus, from AntiSpam to Web Protection. F-Secure Linux Security provides both server and client protection for companies using the Linux environment. It supports a wide varities of platforms. The following 32-bit Linux distributions are supported:

- CentOS 6.0-6.7
- Debian 7.0-7.9
- Oracle Linux 6.6, 6.7 RHCK
- Red Hat Enterprise Linux 6.0-6.7
- SUSE Linux Enterprise Server 11 SP1, SP3, SP4
- Ubuntu 12.04.(1-5) 14.04.(1-3)

The following 64-bit (AMD64/EM64T) distributions are supported:

- CentOS 6.0-6.7, 7.0-7.1
- Debian 7.0-7.9
- Debian 8.0, 8.1
- Oracle Linux 6.6, 6.7 RHCK
- Oracle Linux 7.1 UEK

- RHEL 6.0-6.7, 7.0-7.1
- SUSE Linux Enterprise Server 11 SP1, SP3, SP4
- SUSE Linux Enterprise Server 12
- Ubuntu 12.04.(1-5), 14.04.(1-3)

6. Panda DesktopSecure for Linux

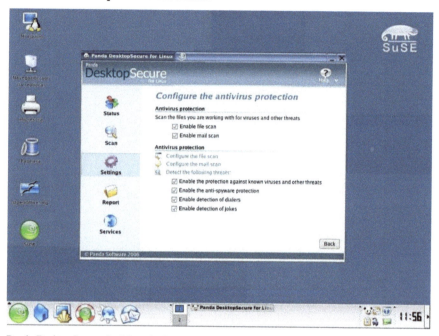

Panda DesktopSecure for Linux is the ideal solution for computers running Linux. It is designed to respond to the specific protection demands of home computers, both laptops and PCs, running this operating system.

This powerful and simple to use antivirus includes a wide range of technologies and a firewall in order to prevent data loss from your PC caused by viruses, hackers and intruders.

- Complete protection – Panda DesktopSecure for Linux neutralizes all types of threats: viruses, worms, Trojans, dialers, spyware, spam; regardless of the source of the threat: Windows or Linux. DesktopSecure also incorporates a heuristic engine that can detect potential threats and block them until the disinfection routine is available.

- Permanent mail protection – Email is the main means of propagation used by malware. It is essential to ensure that you have permanent protection that monitors the email messages sent to users' mailboxes in order to eliminate the threat before it infects the file system. Panda DesktopSecure scans mail reaching the most widely used mail clients, such as Ximian Evolution, Kmail, Mozilla Mail and Thunderbird

- Ease of use – Panda DesktopSecure has been specifically designed for home computers and workstations, providing an intuitive graphic interface based on X-Window, which is fully compatible with the majority of distributions and desktops on the market, such as Gnome and KDE. DesktopSecure can also warn you in the case of incidents and offers clear reports of the scans carried out, and includes a detailed information system to keep you informed.

- Powerful firewall – DesktopSecure for Linux incorporates firewall technology to protect against access attempts from remote computers and external connection attempts from the protected computer, identifying the application involved as if it were a personal firewall. This firewall can also be configured using system rules to administer external connections.

7. Avast Core Security

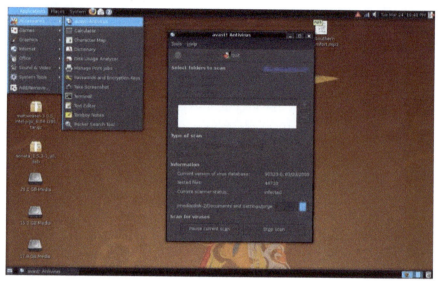

Avast security for linux comes in 3 different package, you can choose to protect your Linux email servers, file servers or the entire network.

- Avast Core Security – Basic antivirus that combat the newest threats and prevent malware from infecting your linux server.

- Avast File Server Security – Protect files right on your server and save the precious power of your workstations for creative work.

- Avast Network Security – A Firewall that scan and filter all web (HTTP) and email (POP3, IMAP) traffic on the network, the filter is completely transparent to users on the network, so there's no impact on your network performance.

8. Rootkit Hunter

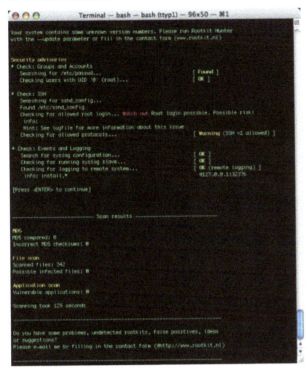

Rootkit Hunter is an excellent choice for Linux rootkit detection. RKH uses the command-line, backdoors, and various exploits. This tool uses SHA-1 hash comparison to detect malicious entries. It's available or Linux and FreeBSD.

Since Rootkit Hunter is written in Bourne shell, it's portable and compatible with most UNIX-based systems.

Features

- Command-line interface
- Rootkit detection
- Portable
- Lightweight

9. Chkrootkit

As the name suggests, Chkrootkit scans for rootkits. This free open-source program operates via a command line interface. It's extremely lightweight, and what's really neat is its usability from a Live CD. Thus, if you have a rescue CD, you may run Chkrootkit straight from that.

The latest release features backdoor and botnet detection, malicious TinyNDS detection, and Linux.Xor.DDoS malware scanning.

Features

- Rootkit detection
- Lightweight
- Can be run from a Live CD
- Command-line interface

10. Bitdefender For Unices

Antimalware Scan - step 3 of 3

	Perform Scan	Resolve Issues	
Results Summary			
Total Scanned Files	320		
Infected Items	0		
Suspected Items	0		

⚡	BitDefender-Antivirus-Scanner-7.6-4.linux-gcc4x.amd64.deb.run.md5	09-Mar-200!
⚡	BitDefender-Antivirus-Scanner-7.6-4.linux-gcc4x.amd64.ipk.run	09-Mar-200!
⚡	BitDefender-Antivirus-Scanner-7.6-4.linux-gcc4x.amd64.ipk.run.md5	09-Mar-200!
⚡	BitDefender-Antivirus-Scanner-7.6-4.linux-gcc4x.amd64.rpm.run	09-Mar-200!
⚡	BitDefender-Antivirus-Scanner-7.6-4.linux-gcc4x.amd64.rpm.run.md5	09-Mar-200!
⚡	BitDefender-Antivirus-Scanner-7.6-4.linux-gcc4x.i586.deb.run	09-Mar-200!

BitDefender does make an excellent Linux antivirus program. But it's decidedly difficult to find on the BitDefender website. BitDefender provides a free trial, and you can request a free license. Unfortunately, end of life has been announced, though downloads are still available. Releases include iterations for Samba (Solaris, Linux, and FreeBSD), as well as Unices (Linux and FreeBSD).

While this tool has reached the end of life support, it's still arguably better to use than nothing. But for a long term solution, it's likely best to use a currently supported antivirus tool for Linux.

Features

- Free
- Installers for Samba and Unices
- End of Life (not supported, though still available)

11. Lynis

Any decent Linux security software will check for rootkit or compromised Linux programs. You can do it manually too: Compare the checksum of programs you have installed with their equivalent on a clean, pristine install system. They should always be bit-for-bit identical. Keeping a system clean is about more than viruses and rootkits. Lynis offers a full set of security auditing tools.

Better yet, Lynis is open source and supports just about every Linux and Unix-based system, including FreeBSD, Linux, NetBSD, and Solaris. It even works with MacOS. If you have existing malware scanners like ClamAV or rkhunter installed, Lynis can automatically tie them into its scans and monitoring, too, checking for configuration errors at the same time.

The entire system is written as a set of shell scripts, not as a block of C++ or something else impenetrable. You can run Lynis directly or install it from a USB thumb drive, CD, or DVD, which also makes it quite portable and a smart addition to any field security specialist toolkit. Indeed, it offers specific guidelines if you need to work on system hardening or compliance testing too, even if your system is isolated from the public internet.

12.   ISPProtect

If you're an internet service provider (ISP), you have a unique set of challenges when it comes to keeping your systems free of malware and policing the files and software that users upload and install. That's what ISPProtect is for. It's quite useful whether you've got dozens of users or a small Linux box in the server rack delivering up Web pages for an intranet.

ISPProtect scans and identifies malware in WordPress, Joomla, Drupal, Magentocommerce, and can also ensure that all elements of these popular third-party software systems are up-to-date. Outdated installations are a common means of penetrating an otherwise secure system.

The program is built around a signature-based scan engine for viruses along with a heuristic scan engine that detects malware in many environments. It can handle lots of scenarios, including spam sent from your server from an unknown software package, an atypically high server load, or even customers complaining about their individual servers. This will make it easy to quickly identify and isolate problems.

One more thing: ISPProtect is written by the same open source development team that created the popular ISPConfig Webhosting Control Panel. An additional part of the package -- ISPProtect BanDaemon -- also protects your system against brute force or denial of service (DoS) attacks.

13. ESET File Security for Linux / FreeBSD

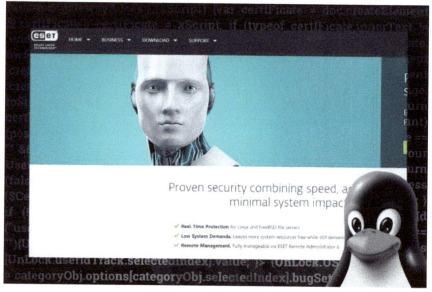

Want a solution from a vendor that covers all the operating systems, offering solutions for Mac, Windows, and Linux? ESET has you covered with its wide range of antivirus and anti-malware tools that include a suite of file security software tools designed to simultaneously keep your Linux and FreeBSD servers clean, safe, and running quickly.

As with many of the other solutions, ESET File Security for Linux / FreeBSD also offers remote management. That's critical if you have anything more than a couple of servers in your facility, especially if you have servers located in offices around the U.S. or globally.

Just as important, you want compliance monitoring to ensure that all the servers throughout the organization comply with company security standards because it's easier to fix it before it's hacked.

ESET File Security works with Suse, Fedora, Mandriva, Red Hat, Ubuntu, Debian, and FreeBSD, offering extensive solution for even the most heterogeneous Linux shop.

14. Kaspersky Anti-Virus for Linux/Endpoint Security for Linux

Kaspersky has long been known as a powerhouse in the antivirus world. In particular, its anti-malware software is popular in the Windows world, which gives the company deep knowledge of malware signatures and profiles, including those on Linux servers.

The company splits its product depending on what kind of systems you have: Kaspersky Anti-Virus for Linux Workstations is designed for an interactive system while Kaspersky Anti-Virus for Linux File Servers is designed for file servers. The company also has a product just for email servers.

With many of these solutions, the question is -- always -- how responsive will the company be to new attacks and exploits. Kaspersky releases database updates every hour, as needed.

How To Install Clamav Antivirus On Ubuntu & Other Linux Distro

Without a doubt, ClamAV is the most popular option for keeping viruses off of your Linux machines and out of your shared directories. There are a few reasons why ClamAV is so popular among the Linux crowd. First, it's open source, which in and of itself is a big win. Second, it's very effective in finding trojans, viruses, malware, and other threats. ClamAV features a multi-threaded scanner daemon that is perfectly suited for mail servers and on-demand scanning.

ClamAV can be run from command line or it with the ClamTK GUI. Both tools are easy to use and very dependable. Installing ClamAV is simple.

For Debian-based systems:

```
sudo apt install clamav
```

For RHEL/CentOS systems:

```
sudo yum install epel-release
sudo yum install clamav
```

For Fedora-based systems:

```
sudo dnf install clamav
```

For SUSE-based systems:

```
sudo zypper in clamav
```

If you're running a Debian-based desktop, you can install ClamTK (the GUI) with the command:

```
sudo apt install clamtk
```

There are also third-party tools that can be added (to include support for the likes of MTA, POP3, Web & FTP, Filesys, MUA, Bindings, and more).

Upon installation, the first thing you'll want to do is update the signatures with the command *sudo freshclam*. Once that completes, you can scan a directory with the command:

```
clamscan -r -i DIRECTORY
```

where DIRECTORY is the location to scan. The -r option means to recursively scan and the -i options means to only print out infected files. If you work with the GUI, it's even easier. From the GUI you can run a scan and, should ClamAV find anything, act on it (Figure 1).

The one caveat to ClamAV is that it does not include real-time scanning. In fact, if you're not using the ClamTK GUI, then to create a scheduled scan, you must make use of crontab. With the ClamTK GUI, you can only set up a schedule for your user home directory.

Biografi Penulis

Cyber Jannah Sakura has been a columnist, health writer, soil scientist, magazine editor, web designer & Aikido martial arts instructor. A writer by day and reader by night, he write fiction and non-fiction books for adult and children.